This book belongs to:

. . . .

MOCHI

With special thanks to
the entire Mochi family

First print edition, Christmas 2019.

www.LearnWithMochi.com

One fine day, Mochi the curious brown BEAR was sailing along the coast of the Mediterranean SEA...

...when he happened upon a MARVELOUS sight.

It was the ancient

3

EGYPTIAN PYRAMIDS!

As Mochi got closer, he realized that pyramids are just 3D TRIANGLES!

A TRIANGLE has 3 sides and 3 angles. There are many types of triangles.

In architecture, TRIANGLES are used to reinforce structures like bridges and buildings.

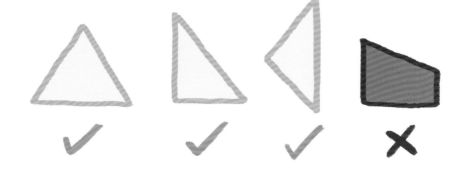

✓ ✓ ✓ ✗

As Mochi got closer to the pyramids, he realized the pyramids were made of thousands of smaller BLOCKS!

each block was like a 3D RECTANGLE.

RECTANGLES have 4 sides.

It is easy to stack rectangular things. like boxes or building blocks.

A square
is a rectangle
of 4 equal sides.

9 ✓ ✓ ✗

Inside the pyramid,
Mochi saw many
SQUARE
chambers.

But MUMMIES were the coolest of all!

When Mochi finally finished the tour, it was very **HOT** outside.

Luckily, Mochi made friends with a **camel**, who led him to a drinking well.

CIRCLES are round and can roll easily.
They are used to roll things, like cars, trains, and bikes.

In the evening, the sun was setting.

Mochi noticed the well's shadow was like a flat circle. A flat circle is an ELLIPSE. ELLIPSES are OVALS, like Mochi's body, a football, an egg, or a watermelon.

Mochi learned a lot
about shapes that day.
Of course, not all objects
are just one shape.

19

How many shapes do you see here in the Great Sphinx of Giza?

More importantly, some shapes are two-dimensional (2D) while other shapes are three-dimensional (3D).

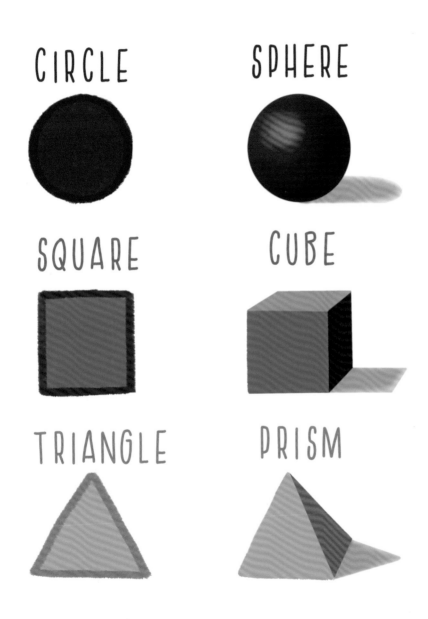

CIRCLE

SPHERE

SQUARE

CUBE

TRIANGLE

PRISM

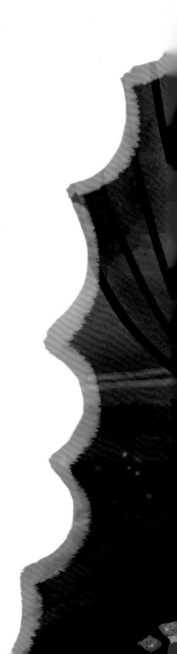

Mochi thanked the camel for being such a great friend on his trip to Ancient Egypt.

"See you next time!" waved the camel, as Mochi waved back.

22

As MOCHI sailed off
into the night,
he dreamed
of many shapes.

Can you name them all?

 **CIRCLE**

 **TRIANGLE**

 **ELLIPSE**

 **RECTANGLE**

 **SQUARE**

 SPHERE

 CUBE

 **PRISM**

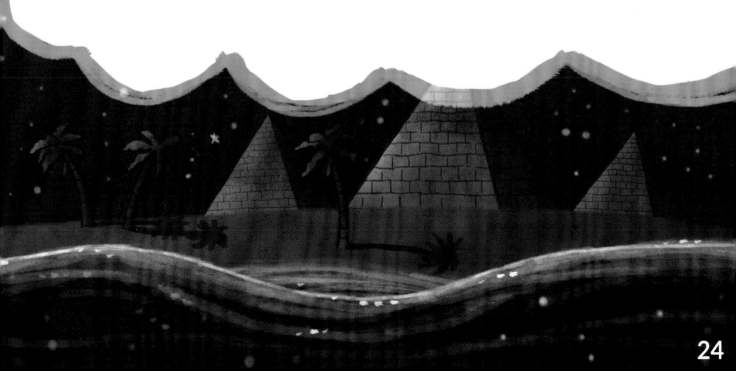

Are You ready?

Prepare Mochi for the adventure!

STEP 1.
Find your wooden
CAMEL accessory.

STEP 2.
Attach it to Mochi's
robotic vehicle.

STEP 3.
You're ready
to explore!

Now it's your turn!

Can you help Mochi go from his SAILBOAT to the PYRAMIDS?

CONCEPT

Algorithmic thinking: creating and following precise instructions

STEP 1.
Find 3 FORWARD blocks,
1 RIGHT turn block

STEP 2.

Place Mochi on the sailboat (e5), facing south at the SPHINX

STEP 3.
Your programming board should look like this. Then press PLEASE GO!

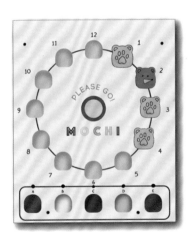

exploring the PYRAMIDS

Mochi is scared! But singing calms him down.
Help Mochi sing before he goes to the biggest
PYRAMID of them all !

CONCEPT

- Parallelization: engaging in simultaneous activities (singing while moving)

STEP 1.

Find **1** SONG block,
1 FORWARD turn block

STEP 2.

Make sure Mochi is on the first
PYRAMID **(C4)**, facing south.

STEP 3.

Your programming board should
look *like* this. Then press
PLEASE GO **!**

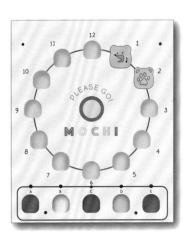

What will You do?

Create your own adventure in EGYPT with MOCHI!

In the story, Mochi befriends a camel and walks to a drinking well after he sees the pyramids. But there are many ways to enact this adventure! Where do you want Mochi to go? Here are some possible paths:

To the camel friend!

To the desert!

To the treasure chest!

CONCEPT

Algorithm design: planning a path by creating a queue

STEP 1.

Decide where Mochi goes next

STEP 2.

Find the blocks you need to go there

STEP 3.

Think about possible routes, plan your path. Then press PLEASE GO!

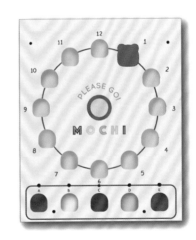

More in this series: